Playful Animals

PRAIRIE DOGS

Ursula Pang

PowerKiDS press

PK Beginners

Prairie dogs are not dogs.

Prairie dogs are squirrels.

Prairie dogs live in a prairie.

A prairie is flat grassland.

Prairie dogs dig holes.

Prairie dogs live in groups.

Prairie dogs watch for trouble.

Prairie dogs bark when they see trouble.

Baby prairie dogs
are cute!

Prairie dogs are playful.

Prairie dogs say hi with a kiss!

Published in 2025 by The Rosen Publishing Group, Inc.
2544 Clinton Street, Buffalo, NY 14224

First Edition

Editor: Greg Roza
Book Design: Michael Flynn

Photo Credits: Cover, p. 1 Roger Utting/Shutterstock.com; p. 3 Peter Krejzl/Shutterstock.com; p. 5 Marie Dirgova/Shutterstock.com; p. 7 Becca in Colorado/Shutterstock.com; p. 9 Guy William/ Shutterstock.com; p. 11 Poly Liss/Shutterstock.com; p. 13 Jaren Jai Wicklund/Shutterstock.com; p. 15 l i g h t p o e t/Shutterstock.com; p. 17 Wirestock Creators/Shutterstock.com; p. 19 Georgi Baird/Shutterstock.com; p. 21 inge-klikt/Shutterstock.com; p. 23 Nick Fox/Shutterstock.com.

Library of Congress Cataloging-in-Publication Data

Names: Pang, Ursula, author.
Title: Prairie dogs / Ursula Pang.
Description: [Buffalo] : PowerKids Press, [2025] | Series: Playful animals
Identifiers: LCCN 2024031258 (print) | LCCN 2024031259 (ebook) | ISBN 9781499450804 (library binding) | ISBN 9781499450798 (paperback) | ISBN 9781499450811 (ebook)
Subjects: LCSH: Prairie dogs–Juvenile literature.
Classification: LCC QL737.R68 P36 2025 (print) | LCC QL737.R68 (ebook) | DDC 599.36/7–dc23/eng/20230412
LC record available at https://lccn.loc.gov/2024031258
LC ebook record available at https://lccn.loc.gov/2024031259

Manufactured in the United States of America

CPSIA Compliance Information: Batch #CWPK25. For further information contact Rosen Publishing at 1-800-237-9932.